An Ethic For Our Time

AN ETHIC FOR OUR TIME

AVRIL FOX

authorHOUSE®

AuthorHouse™
1663 Liberty Drive
Bloomington, IN 47403
www.authorhouse.com
Phone: 1-800-839-8640

Published by AuthorHouse   09/17/2012

ISBN: 978-1-4685-8325-0 (sc)
ISBN: 978-1-4685-8326-7 (e)

This book is dedicated to Sir David Attenborough, who, like myself, is a member of the Optimum Population Trust.

Contents

For want of a nail, a shoe was lost,
For want of a shoe, a horse was lost,
For want of a horse, a rider was lost,
For want of a rider, a battle was lost,
For want of a battle a kingdom was lost,
And all for the want of a horse-shoe nail!

This is a nursery rhyme, repeated by women to their children down the ages, part of our culture, which is disappearing. It is also part of an ancient wisdom, which we are losing. We have so many of these phrases in our culture which—contrary to our present philosophy—we are forgetting: "a stitch in time saves nine"; "waste not want not" (even more important today). All British parents who have not done so should buy a book of nursery rhymes—preferably by a person called Opie—and start to read them aloud to their children: this is an important part of introducing them to books. We think we are so clever in our age, but we should re-learn these ancient wisdoms.

Science is really only the discovery of Nature's laws.

Even in the Christian Old Testament there is a whole book called "*Proverbs*".

An Ethic For Our Time

Introduction

I am very old, and the world seems to be blind. Nobody appears to see that we are killing the ecological world we depend upon: the tiny creatures, and the great predators: the tiger, the Polar bear, the koala, the lion. Then there is the system: who chose capitalism? Seems it "just grew" and it is leading us the wrong way. The Market has become a sacred cow, but there is a greater factor, which has been forgotten: it is Nature that should dominate, and its values: they should have priority over all. Then harmony would rule.

I was brought up in the Cotswolds, a beautiful area. My father worked for a garage, and one day took us out for a run in a car (in those days only rich people had cars). Suddenly I shouted out "But are they allowed to do that?" "Do what?" asked my Mother. I could not explain; I was too young—but I now know what I meant: "Build horrid, ugly buildings!" I had never seen ugliness before, and was just shocked by it. Because I was really horrified, but did not know why, that memory has stuck in my mind.

Now, in his book just published: Religion for Atheists, *Alain de Botton* suggests that we think up a substitute for religion. While I sympathise with this, I realise that from that day when I was so startled at ugliness when I first sighted it, my whole life has been motivated by that moment: horror at the loveliness destroyed by humanity. But in order to discuss that, let us go right back to the beginning: the stage when life began: the time of the hunter-gatherers. At this juncture we enter prehistory.*

They had a religion which has never been improved upon, a goddess recorded in words I cannot better: the words of Ernle Bradford, in his book Mediterranean: Portrait of a Sea: *"her groves were in the high places, and her worship is the worship of life itself. She remains for all time the goddess of the Mediterranean peoples. The most ancient of deities, the fertility goddess originally dominated the whole sea. In the Near East and the Aegean her cult had long been paramount. In Crete, the Minoans had paid their tribute to her as a lunar and a snake goddess, and also as mistress of the beasts, the Lady of Wild Things. She had been worshipped in the Maltese islands since the Neolithic period. Fecund and tranquil, her steatopygous images invited men to produce their own kind, to cause their livestock to do the same, and to ensure that the fields grew fertile. She was a peaceful goddess. In the Maltese archipelago, during her uninterrupted reign of nearly*

---

*   Publisher Hamish Hamilton

one thousand years, there was no warfare between the agricultural settlements. No weapons of this period have been found and, even more significant, there are no signs of man-made destruction in any of the temples devoted to her cult."

It still lives, and has simply one fault: self-contradiction. It is "the Mary cult", and it lives on. One of its strongest protagonists is Geoffrey Ashe, and the title of one of his books betrays the weakness: The Virgin. *A virgin cannot be a mother; this is the difficulty.*

But there is another source of information. I spent many hours years ago, before it moved out, in the British Library, searching for the source of the insistence of the People of the Book (Jews, Christians and Muslims) in distaste for sex, and I have got no further than the Orphics, which aroused the hatred of Greek women, as portrayed on vases hundreds of years ago, but no details about it. There is nothing I could find apart from these.

Before this concept took hold, in the time of hunter-gatherers, Neolithic communities held men and women as equal: the men hunted and the women equally, children and the old and infirm: gathered. When a decision had to be made, the Wise Council got together (in old English something of it still lived: the Anglo-Saxon Witenagemot*), men and women, and came to a conclusion.*

The History *of Christianity, by Diarmaid MacCulloch*[*]*, in its early pages, describes a "neurosis"*

---

[*]   ISBN 978-0141-02189-8

of history: regret for a lost golden age: "a moment of history when all was well". I believe in this, and with good reason. In the same book we are told that the Gospels of the New Testament finally were bought together in 367 CE: there was plenty of time for garbling of the message by that date. I do not believe Mary was celibate after her marriage to Joseph; Jesus had brothers and sisters later. I believe that Jesus existed, and that he was a great preacher, ahead of his time

But something had changed: religion itself. There had been a number of new ones, each led by a prophet, competing for converts. Among them was Christianity and Buddhism, a fairly old religion. They preached reincarnation, which at that time was believed by most religions. The Christians decided they would be different; they would say that, as it tied in with the old religions, certainly there was a Saviour who gave his life for us, only he had died on the Cross, and redeemed all the sins we had committed in the only life we had. This was very attractive, as everyone had some sort of sin. But in taking this new course, Christianity lost the humbling aspect of reincarnation: if you had only lived once, you lost the knowledge that though you might be a Prince, or a rich man, in this incarnation, you might be a beggar in the next!

Our concepts of religion began in the Mediterranean, and here I turn to a source that has never been challenged: Robert Graves, in his unsurpassed Greek Myths, *first published in 1955, by Penguin Books*[*]. He

---

[*] ISBN 014-02-0508X

has never changed the piece I quote in any edition. He writes (abridged): "A study of Greek Mythology should begin with a consideration of what political and religious systems existed in Europe before the arrival of Aryan invaders from the distant North and East, The whole of Neolithic Europe, to judge from surviving artefacts and myths, had a remarkably homogeneous system of religious ideas, based on worship of the many-sided Mother-goddess who was also known in Syria and Libya. The Great Goddess was regarded as immortal, changeless, and omnipotent, and the concept of fatherhood had not been introduced into religious thought. Then, when the relevance of coition to child—bearing had been officially[*] admitted, man's religious status gradually improved. The tribal Nymph, it seems, chose an annual lover from her entourage of young men, a king to be sacrificed when the year ended, making him a symbol of fertility. From the day he was chosen he was honoured and cherished (the bodies preserved in the Celtic bogs reveal this in their manicured nails and other signs). His sprinkled blood served to fructify trees, crops and flocks, and his flesh was torn and eaten raw by the Queen's fellow-nymphs-priestesses wearing the masks of bitches, mares or sows. There is, however, no evidence that, even when they were sovereign in religious matters, men were denied fields in which they might act without female supervision: they could be trusted to hunt, fish, gather certain foods, mind flocks and herds, and help defend

---

[*]   ISBN 014-02-0508X

the tribal territory against intruders, so long as they did not transgress matriarchal law."

This might well be the source of the legendary "golden age": suffice it to say that it was peaceful, which the people of that area remembered nostalgically when the Aryan gods and male rulers took over.

I was interested in the account by two brothers who were studying the Chukotka tribe in Siberia, according to a Channel 4 TV programme that I watched on the 5th February 2008. They interested me because their diet consisted largely of meat: especially walrus and whale meat, but their cholesterol and general health was superb. They are Stone Age people, but have permission to cull 50 walruses when the animals arrive in summer, so that they can store them over winter. But they particularly attracted my attention because when they hunted and killed an animal they do it with spears because guns upset the whole herd. Once it is dead they stroke it and thank it on behalf of the ancestors, in a manner which the brothers found intensely moving.

On quite a different subject, there is an ironical misunderstanding in Herodotus *in this connection: in* Book One, *he cites it as "wholly shameful": that the Babylonian women "who are natives of the country must once in her life go and sit in the temple of Aphrodite and there give herself to a strange man . . . when she has lain with him, her duty to the goddess is discharged . . . after which it will be impossible to seduce her by any offer". He is puzzled by this, but it is motivated by devotion to the Goddess of Love—in the same manner in a truly Christian society a devoted Christian woman would do the same thing at least once, so that no man could go uncomforted, however unattractive he might be. I have searched in vain for the origin of the change to sexuality revealed by Herodotus' use of the word "shameful"; at some time before he lived the view of sex changed, and I regard this change as a retrograde, backward step by the human race. I long for a time when sex takes its place as a sacred, exquisite pleasure to be enjoyed, but only intentionally, and rarely, for procreation.*

So far as we know, sex has been regarded by the people of the Book (Jews, Christians and Muslims) in this new way—as something "shameful", exemplified by the Christian "original sin"—ever since. Archaeology may reveal how it came about: (Northern gods again?) but this step was deplorable.

The invasion of the Mediterranean by northern peoples, among them the ancestors of the Dorian and Ionian Greeks, changed the pattern of life, says Ernle Bradford. The newcomers brought with them a belligerent sky-god, with lightning in his hand, and thus ushered in the Warrior Age. This was a calamity for the people of the Mediterranean, as well as the British, so far as we know. I have read that on the pillars of Stonehenge there are carved certain Mediterranean-type signs, and no doubt we at one time shared Mediterranean culture—the Vikings got everywhere. But the new belligerent pantheon took over, and peace became a folk memory. There is very little record outside Europe of these times, but I hope that one day we shall learn something from archaeology. The new gods were primitive and barbaric, yet we still call the days of the week by their names: Tues' day, Woden's day, Thor's day, Freya's day, the day sacred to the Sun, the day sacred to the Moon.

Since then all religions have been introduced by male prophets, and all have included male gods and have been warlike. Despite Christ's teaching, later on in history, Christians did not prove more peaceful. The year I was born, 1917, Britain was killing thousands of other Christians in Germany, and the guns were being blessed by a bishop of the Christian Church. They don't go so far as that now. which is a slight improvement.

In this book I suggest that there is no need for yet another man-made religion: we cannot do better than revert to the age-old Lady of the Wild Things, under whom her worshippers, "simple Stone Age people," developed farming. She was, in fact, Nature personified; the people of that time talked of gods, giving them personalities according to the natural laws. In fact the "science" of today is simply Nature, her laws now better understood by humanity. (There are more truths to be discovered, for example: dowsing.)

My lifelong passion is Nature, and what we are doing to it, which I sensed in my childish way that day in the car. I will also suggest that we are ignoring "the elephant in the room" while we discuss endlessly the passing worries of the day, and it is time we noticed it: the human population, and what it is doing to the Earth and to the ecological laws.

One more point: the law of connection: all things connect. I shall mention this later, but bear it in mind, it is one of the laws of wisdom we once knew, but have forgotten.

The spiral of history has moved on, coming to a stage where we are in the same place we had reached, more than a thousand years ago, but on a higher level of understanding, and thus fulfilling an ancient wisdom.

Robert Graves has written of this in his prescient book The White Goddess,* *as a symbol of death in life, regularly re-occurring ("our king has gone to Spiral Castle" meant "our king is dead"; a revolving wheel before the door of a castle was common, and in front of a doorway in New Grange, a prehistoric site in Ireland, there is a broad slab of stone carved in spirals; I have seen the same thing, also carved in stone, in South America.) The spiral symbol is as old as humanity itself. On the old religions, Graves says "Arianrhod is one more aspect of Ceridwen, or Cerridwen, the White Goddess of Life-in-Death and Death-in-Life, and to be in the Castle of Arianrhod is to be in a royal purgatory awaiting resurrection. For in primitive European belief it was only kings, chieftains, and poets, or magicians, who were "privileged to be reborn". Word of mouth changed religions like this as they passed from country to country, but the old wisdom remained. We did not have the written word, and when we did it was only locally. But we thought, and discussed, and the local wisdom was passed down for generations, passed on in rhymes of children, and in the songs of the bards, which are also described by Graves.*

Now we have reached this stage we need a new ethic, based upon certain principles. Whether we call it a religion is unimportant; humanity has perhaps reached a stage in which it is almost adolescent, and no longer requires the mental eiderdown of a God. Indeed, unless we do reach maturity very soon, it will be too late; we are killing off natural ecology already.

---

*   Publisher Faber and Faber Limited, 2nd Edition

A final introductory word: I write not as a learned, academic person, but for the common citizen, (the man in the Clapham Omnibus) for this is a vital message which must reach everybody. Otherwise we shall go, unheeding, into the abyss of oblivion, with the rest of life, and lose the whole precious layer of Life on Earth.

Chapter 1

The Human Race

Thinking in geological time, tectonic plates, at first clustered together in Pangaea, separated, and are continually wandering. Each one developed its own forms of life, according to the conditions there and how close it was to its neighbours before they drifted away. There was one exception: Australia, which did not in relatively recent time get close enough to any neighbour, and as a result its animals stayed for ever in the marsupial form.

Life began to develop first in the oceans, then it crept on to the land, and the dinosaurs developed, but there was an upheaval in the Universe, and they disappeared, but Life recovered, grew wings and found a new dimension: the air. There was now enough oxygen made by the plants, led by grass, for this to be colonised, in an area near the Earth.

Now we have to stop thinking in geological time and think in human minutes, hours and years. Nearly 200,000 years ago types of human beings began to appear, all in the continent of Africa. The last of these has named itself *Homo sapiens*.

Let us establish one fact at the beginning: *there is only one human race*. What we call "racism" is simple colour prejudice. The world has become so crowded that we now know that there are black, yellow and "red" humans (red Indians being a myth created when Christopher Columbus thought he had gone round the world as far as India! For some reason it has persisted.)

The first aim is the most difficult, for it is against natural law: each species wants above everything to *increase* its numbers. But we have to do the opposite, after all we are breaking the laws of Nature every day: we pollute the pure ocean from our ships, throwing our rubbish, largely plastic cups, into it. We have learnt to behave very badly: let us for once learn to break natural laws beneficently! As it is, it will take us years to undo the harm we have done by making it difficult for the little creatures (so essential to ecology) to build their shells, because of our carelessness. Let us learn the first law of Nature: make sure of the scientific results that will follow if you depart from the natural law.

So, first of all we have to reduce our numbers, and that is going to be very difficult. Nevertheless we must do it, because we humans have become a burden upon Nature; for the first time she is being affected by our activities. We are already attacking the large predators, reducing their gene pools to a dangerously low level. Even more serious is the extinction of the tiny creatures upon which Nature depends—the caterpillars which feed nestlings which will grow into our songbirds; in fact, we affect Nature in too many ways to describe here.

Secondly: we have to raise the standard of downtrodden women in half the world, where those who bear the children of the future are treated worse than dogs. Our women in the West have not yet achieved equality, but their situation is improving: they are educated and on the way to equality. I do not want a matriarchy—that would be equally unsatisfactory: we simply demand equality. Woman and man make the natural superbly balanced, pair.

Thirdly, we are at present dominated by cultures, which are mostly based on nothing but superstitions and mistaken convictions. They are very powerful: our culture is developed from the first day we live, but we have to scrutinise it critically and make it ethical, human and—above all—kindly. Take diet: it is natural that humans eat animals; if they did not the balance of Nature would be changed. But in Stone Age days we humbly thanked the Great Spirit of the creatures after a kill, a habit which we would do well to bring into the present. And we should bring back hunting: that is fair, and healthy for the species. The British TV presenter Hugh Fearnley-Whittingstall provides our viewers with the healthy way to find food.

An event in history, which has worsened the relations between nations, is the fact that the relatively-newly-formed country of the USA has decided, on the discovery of nuclear science, that certain nations only have the right to hold nuclear weapons. No nation, which was mature, would have done such an unwise thing: far better to say "no nation will hold nuclear weapons" and make them illegal for all. It follows in this man-dominated world, that if there is one nation which holds this right, others will wish to do so; man dominating alone is always competitive. Only a nation which is insecure, and not "the First People"* in their land, would behave thus. As a result we now have a world in which this is a matter of contention, which was quite unnecessary.

A wise human race would change the condition of the world gradually but permanently, and for the better. It can only be one which women are equal with men, for there are two kinds of wisdom: male and female (there are arguments about this, of which I am aware, but I am certain about it; I have tested it by travelling round the world and studying many cultures.)

It will mean a gargantuan struggle; at present we settle our differences by warfare, which is negative, for the strongest always wins, not necessarily the one who is right; a new human race will emerge: civilised and kindly, educated and caring, fit to be the parents of the future, and in comfortably-small numbers, so that the wild things have space in which to live naturally.

Those who agree with me will be called the Realists, because we see existence as it really is, with no illusions. We shall soon be clearly distinguished from artists or other types of realists because of our opinions: reducing the number of humans on Earth without slaughter; raising the status of women;

---

* A phrase coined by David Maybury-Lewis in his book 'Millennium'. (Viking)

peace-loving; argumentative but always rational. Unlike the present movement, which is vaguely unsatisfied with capitalism, we are not vague at all: we know precisely what is wrong with the state of the world, including capitalism and its greed for profit above everything else. The basic ethic of a Realist is ecological: to strive to return to a balance with Nature. This undoing of the harm humanity has done to Nature will take at least a hundred years, but with the men and women of the world together we can do it.

One of the Realists (though he is not yet aware of this!) is Bruce Parry. He has drawn together the indigenous peoples of South America and backed them in their struggle to keep unspoilt the great forests of the Amazon; alas, a great deal of these lungs of the Earth have already been lost, but they are now beginning to be saved. He goes back to these tribes from time to time, and each time we learn more of how their struggle is developing. Bruce Parry deserves a special chapter in the struggle for the future: in his last visit to the various tribes he goes to a gathering of the Kayapos, is welcomed, painted in celebration and celebrates his arrival. Then we hear of the black side of the Brazilian Government in the struggle to destroy these forests: illegal slavery of thousands of people from various sources who are lost souls (plenty of these in South America) working in terrible conditions, in which they illegally fell the trees of the forest, which are so precious to the world. Among these are the Brazil Nut Trees, which I saw when I was there: these trees are especially valuable, because they cannot be transplanted into plantations elsewhere, and have to grow to maturity before they can produce those delicious nuts which we eat around Christmas. They then grow into the great nuts, which are harvested and opened to reveal the little ones inside, with which we are so familiar. These are stored in sacks and taken to boats, and are then exported abroad. I remember how I watched this process, tasting one of the fresh nuts, when they filled the sacks. These trees are now felled in the clearance, and are becoming increasingly rare. Please take note of this and join the struggle to save the forests of Brazil, and the First People who are fighting for their survival as tribes. Above all, support Bruce Parry in the work he is doing.

Chapter 2

The Degradation
of Women

This chapter has to be a long one, because I have to bring home to my readers the suffering, which the women of roughly half the world have to undergo. It is better in the West; although we have made significant progress, we still have not gained equal status. In this country we have achieved a woman Prime Minister, but one only has to watch international gatherings on television to see where the real power lies! We now have to win this struggle: it is very difficult, because we have to deal with an undeclared adversary: the fear of the men of what will happen if we allow women equality with them: it goes deep, and is instinctive. Yet there is nothing so comforting and strengthening for a man as to have a strong woman at his side in time of trouble.

Today, one sex in roughly half the world has been degraded. It was not always so: thousands of years ago, in the world where population was still on a low level and traveller's tales told us all we knew of distant lands, it was different, so far as we know. We lived off a bountiful Nature, humans were rare, and everybody had enough to eat. Even when we thought the world was flat, we travelled in canoes—and it was remarkable how far we sometimes managed to go then—and we saw only neighbouring countries, human beings probably lived more or less equally. We don't know how; one day archaeology will perhaps tell us, if we ask the right questions.

In the Christian Bible story, Eve was made responsible for the Fall of Man. That explains it to Christians. But there is another, far older story, told in the *Introduction*: of the original Great Goddess, who brought the peaceful world which humanity remembers nostalgically. The details of this world are now lost to us, one day to be rediscovered by the archaeologists in detail, when we will learn how they came about. Somehow the status of women was degraded everywhere, except—so far as we know—in Crete, where the Minoans thought highly of the female, and in Etruria, where the women were so free that it scandalised the people in neighbouring countries, and all their archives were destroyed by their successors.

Those who came after brought the new, sky-gods, brandishing thunder and lightning. I believe there were other lands where the same story applied, but we know very little of the early days outside Europe. Ernle Bradford suggests that it was when the patriarchal, sky-god peoples invaded the Mediterranean, and the

Warrior Age began, contrasting to the peaceful time, which reigned when people worshipped the Great Goddess: another name for Nature.

Imagine that there had existed no degradation, and we still lived in a state of equality of the sexes, woman and men making the decisions together, things would be very different. For example, there would be no wars: women hate war, it means the bully wins, and children are killed. We therefore would not have conquered the colonies with guns, we would have traded with them, and built up mutually profitable agreements, so that there were plenty of jobs for people in both countries.

Today much goes on unnoticed even now. David Aaronovitch in *The Times* of 15th March 2012 remarks how three women were raped at different time and didn't report it to their parents or the police—I too, was, once, but I was already a married woman; I too did not report it.

I think that there would have been another difference: society would not have expected the impossible from women. On the 14th September 2011 the newspapers announced, in one fashion or another, the report of UNICEF on the state of the family in European countries. Britain received a low mark; in the report on the same subject four years earlier Britain was marked even worse. We are a crowded island, and not usually a particularly cruel nation, but in the contemporary world the lot of mothers and children is growing to be very difficult, for the workload is great.

I think the time has come when women have to think, as they reach puberty, whether they wish to be mothers. If they do, they must decide that for at least ten years after a child is born they will devote their life to its upbringing. Only connect: such thinking harmonises with our realisation that the human burden on the planet is too great: we shall have to keep our population down.

Also, not every woman wishes to make her children her life's aim. Many would rather dedicate their lives to art, or the stage, or other human beings—even the children of others, adopted. When it comes to adult life, parenting should be a choice, not everyone is born to it or even very good at it. In *The Times* of 11th of May, 2012, under the heading "Mothers want More Help", we learn that two in five new mothers during the first five weeks after birth were very upset at their child crying, according to a survey done for the NSPCC by YouGov. Almost three-quarters wanted advice on how to deal with anxiety and depression.

It is time we in Europe turn to an old custom which was a very good one: when puberty comes, children should learn about serious things: finance, nutrition, sex. For a year they should be separated from the community, and each other, to learn about these. In the case of sex they will learn that sex results in babies unless one takes precautions; it can be ecstatic in puberty, when the hormones are at their highest peak, but they should learn how to avoid procreation, and how to enjoy sex without it. But it is essential that they learn how tedious it is to rear unwanted children, so that Reality is rammed home.

These realisations came to me in a roundabout way, and much thought. In my travels I visited Santorini, a Greek island half of which was once destroyed by an earthquake so huge that in the year 535 AD (CE) in distant England the sky darkened, and the weather was not normal for two years because of the spread of volcanic ash, though this was not known at the time. (It is reported in the *Anglo-Saxon Chronicle.*[*]) When I visited the island I saw the archaeological dig that was proceeding at Akrotiri, and noticed that the Minoan women were portrayed in two different ways in the art work being revealed: some of them had breasts that were firm and the others had breasts which hung low when they bent over. This was so noticeable that it made me ponder: it signified some difference. On consideration I realised it made clear that there was a difference between women who were virgins and those who were mothers. Many years later I came to the conclusion that this difference was significant, and that they were ahead of the women of today: they realised that they had to decide, at a certain time in their lives, whether they would become mothers, or remain free for other things.

This applies even more today: if we were wise, the modern woman, at some stage, would face a similar situation, but modern culture does not allow us to make the choice. It comes at the stage when we take on a permanent relationship with a man: then the Minoan woman decided that she would or would not have children. (I wonder how they carried this out? Either they practised some sort of birth control, or they made sure the infant did not live.) If she did not, her breasts would remain girlish, if she did, she would one day suckle a child and they would change their shape.

---

[*]   ISBN 434-98210-5

If in our day we recognised this fact of life it would save women much heartbreak. As it is, they take on two jobs, mother and worker. The child suffers if there is not the constant attention of a mother throughout, say, ten years, between mother and child; the mother suffers because she is constantly torn between her duties at home and her work. If she made a choice at the outset of her marriage, she would either have no children and a satisfactory career, or have them and know the joys of motherhood. She may change her mind later, but the decision rests between the husband and wife.

If we took on a cultural habit of making this decision, as, perhaps, did the Minoan and Etruscan women, it would be the beginning of the drop in population which is now so desirable; if we do not it means more suffering for children, wives and husbands, and eventually, if we do nothing, the destruction of the ecological web upon which the whole of life depends.

A rise in the status of women would have saved us also from the present migratory drift of people, with a sad litter of starvation and death along the route. And there would have been no need for the colonies to rise and fight for their freedom—it would not have arisen; in fact history would have been altogether different. The rise of Islam could have occurred, but it would have been different in character—for it would have taken place in countries where women were educated.

An example taken from Egypt in 590 BC is particularly significant in this respect: in that period Egyptian women were very strong and highly valued in the community, as can be seen from this excerpt from the marriage agreement between a man and a woman:[*] "I acknowledge thy right as a wife, from this day forward I shall never by any word oppose thy claims. I shall acknowledge thee before anyone as my wife, but I have no power to say to thee 'Thou art my wife', it is I who am the man who is thy husband. From the day I become thy husband I cannot oppose thee, in whatsoever place thou mayest please to go . . . I have no power to interfere in any transaction made by thee."

What would any woman in Asia not give for such a marriage contract today, or even a European woman? It would be out of the question. We do not know what occurred in Egypt between 590BC and now, but the wording of this contract indicates that there was already some struggle which needed to be emphasized. It is sufficient to say that this existed then, and does not exist now; then the Romans came to Egypt, and destroyed all the temples, replacing them with those acknowledging Caesar. The Egyptians did not discriminate between religion and life in general, so that all their wisdom and their past history was destroyed with the temples. We have snatches of wisdom about Hermes Trismegisthus, but just enough to whet our appetite, and that's all.

I think that the degradation of women would not have happened if the sky-god worshippers had not overwhelmed the Mediterranean, and killed off the peaceful people of the Great Goddess: in the course of time archaeology can discover the truth (but you have to know what to look for). I suggest that at first

---

[*] Publisher Volturna Press, 1971(ref.'The Mothers' by R. Briffault). Also in 'The New Matriarchy', by Evelyn Acworth.

in prehistory women were held to be sacred, in some countries at least, perhaps in all, as in the time of the Great Goddess. They bore the children, in the women rested the future of the tribe. They laid out the dead, in Greek mythology the Three Fates: Clotho, Lachesis and Atropos, were female. When humanity lived in the hunter-gatherer stage, the powers of Nature were seen as the various gods, and were given names under the Goddess, who personified the Earth. The belief in a Father God, wielding thunderbolts, was a product of the Warrior Age, one of patriarchal gods. The Sibyls of the days before this were always women; their oracles were bought by those practical people, the Romans, from the Greeks, they made pronouncements which were pored over, word by word, by governments and politicians in the ancient world within the ambience of Greece, which was large and its culture strong.

Written history did not start until after thousands more years had passed; by which time women in many parts of the world had become downtrodden, uneducated and, if married, continually pregnant (if they were fertile). In some cultures there were harems, in which chosen women were segregated under the care of eunuchs: men who had been castrated and thus could not impregnate them.

In the world of today, women are unequal still. In Asia they are utterly degraded: as will be seen. In Europe they are more equal, but men are still in positions of influence; in South America and elsewhere their position varies, but only in Scandinavia are they equal to men.

I have kept a file for a few years on the indignities suffered by my sex in Asia: it is a very fat one. Here are a few, all taken from *The Times*, starting on the 17th October 2009: when women in Mogadishu, Somalia, were whipped in public by al-Shabah militants for wearing bras—part of a movement for enforcing Sharia laws.

The next item, dated 4th November, 2009, comes from the land of the free, the USA:Atlanta, Georgia: a 20-year-old Iraqi woman living in America died after her father allegedly ran her over because she was "too Westernised"; the police report said Faleh Hassan Amaleki drove into his daughter Noor and her boy-friend's mother, then tried to flee to London, but was refused entry and arrested on his return to Atlanta.

This is followed by a lengthy item dated 28th November 2009, from Ampatuan in the Philippines, concerning a man called Ismael Mangudadatu, in the Phillippines, who wanted to stand for election but knowing his enemies were powerful, if he appeared in public saying so he would be asking for trouble, so he consulted with his family and sent his wife and her two sisters, along with the two family lawyers (both women) and an aunt (who like one of the sisters was pregnant), to register. As a further safeguard, 27 journalists were invited to join the party: "they will not be harmed if journalists are watching". At 9.30 a.m. they all set out by bus from the town of Buluan to the provincial capital Shariff Aguak. Within an hour the wife and her companions were all dead. They had been bound, shot and buried in a mass grave—it is believed that some women were raped. This appears to be part of a struggle between two family gangs in the Philippines. Three journalists survived the attack because they had had a warning from a motorcyclist with a gun asking for the names of journalists on the party. Police Chief Felicisimo Khu said in his report on the mass grave: "They were buried in six layers. Every layer is covered with earth making it more difficult for retrieval . . . one of the party has no underwear.". "To judge from the blood, some of the corpses may have been buried while still alive," said Bemito Molina, a forensic examiner at the scene.

Our next is from Turkey on the 18th December 2009 about a teenage girl who was murdered by her father to protect his family's honour; the family pleaded with the father, Mehmet Goren, to reveal where her body was hidden so that they could give her a proper burial, and thus achieved the headlines. Goren's defence was that he killed his daughter after she rejected the customs of her homeland, embraced Western life and formed a relationship, bringing what he saw as disgrace on him and the rest of his family. He was convicted and told that he would serve at least 22 years in prison. It was the tears of the sister, Nuray Guler and the jury that drew the headlines. The family are worried about the publicity: in her statement Mrs Guler said that her mother might suffer because she had given evidence against her husband: "No-one should fail to realise what this means within our culture—she had made a statement against the men of the family". We in the West do not realise how powerful culture is, although we too follow our own cultures just as strongly—fortunately they follow the Anglo-Saxon pattern.

A similar case dated 31st December comes from Amman in Jordan, where a man aged 61 was sentenced to ten years' imprisonment for killing his 17-year-old daughter. She had been kidnapped and raped by a group of men, who are now on trial, but it was culture that killed her. The men were not punished.

We go to Mexico to an item that deals with a campaigner for women's rights in that state: she has drawn attention to the murder of hundreds of women around Ciudad Juarez, on Mexico's border with the USA. Their bodies are being dumped in the desert around the city or left in wasteland. The series of killings began in the early 1990s, and have been treated with indifference by the authorities.

Her name is Esther Chavez, and she is the founder of Casa Amiga, the rape crisis movement, which has now some 50 of its centres throughout the country, sheltering young girls from the countryside who are attracted by the booming industries on the border states. They have no families to support them when they supply cheap labour for the factories, which are US brands, using imported materials which they export after they have been processed. Chavez shamed the government into prosecuting some people for the sexual offences, but then found they often picked on innocent men and came to the conclusion that the killings were not the work of a "serial killer" but were due to a more complex problem: a country that values "machismo" resents women gaining employment in textile plants, which gives them independence, and some men began violence in return. The term *feminicidio* was coined to describe what resulted; others believed that drug-trafficking gangs were involved; and Chavez herself believed that the state police, judicial officials and business leaders were themselves implicated in some of these murders.

Esther Chavez was born on 2nd June, 1933; she died of cancer on 25th December, 2009. She is an example to all who care about the welfare of women in that half of the planet where they suffer. A film has been made about her.

Our police force in Britain does not emerge from the record as faultless: a news item dated 20th January 2010 reported that when a woman complained that she had been sexually assaulted by a taxi driver, said that the policemen laughed when she listed her injuries, she was lied to, made to feel like a criminal rather than a victim, and patronised. Had they paid proper attention to her complaints, other women would not have suffered from the same taxi driver. As it was, it took some time before the driver was apprehended: he was English, he had carefully planned all his assaults, drugging the women beforehand—and he drove a black cab, not a minicab. The key fact is the police did not pass the case to the Crown Prosecution Service, as they had promised her—in fact they had taken the man as telling the truth. His name was John Worboys: he was finally convicted in 2008, when he had achieved further victims; the attack took place in summer 2007.

A horrifying story comes from Turkey on 5th of February 2010 of a father and grandfather of a teenager who are to face trial for burying her alive because they were concerned that her friendship with boys had brought dishonour on their families. Medine Memi had been discovered, bound and lifeless in sitting position in a 2-metre hole dug beneath a chicken coop outside the family home, 40 days after she had disappeared: the hole had been cemented over. A post-mortem examination revealed a large amount of soil in her lungs and stomach showed that she had been buried when conscious and suffered a slow and agonising death. It also emerged that Medine had repeatedly tried to report to police that she had been beaten by her father and grandfather days before she was killed. "She tried to take refuge in the police station three times, and she was sent home three times," said her mother, Immihan, after the body was discovered. No doubt the sensational nature of the murder enabled it to get publicity: honour killings of this kind occur mainly in the Kurdish south-east of Turkey. Once again, the traditionality of a culture does not guarantee its virtue: if the culture harms the weaker section of the population it cannot be justified. I am increasingly convinced that anti-woman attitude dates back thousands of years.

Our next item comes from Saudi Arabia, where the women are now struggling for the right to drive a car, but have in 2011 been given the right to vote in elections. It is dated 9th February 2010, about a 12-year-old girl fighting for a divorce from her 80-year-old husband. The state-run Human Rights Commission has hired a lawyer to represent the girl when she takes the case to court in Buraidah, a conservative town near the capital Riyadh; the Government had taken a different course on a previous case. On the 22nd April 2010 we see a report that she won her case: a step forward in a fight which should be international in character.

This one is, and still continues—it is worth noting that Navi Pillay, the UN representative, is herself a woman. In comments on the above-mentioned case, *The Times* notes that: in April 2009 a Saudi judge refused to annul a marriage of this kind because she had not yet reached puberty! The anecdotal evidence suggested that in Saudi Arabia child marriage for females was high; it became a *cause celebre* in 2008 when Nujood Ali, 10, sought divorce from an abusive husband more than 3 times her age. Marriage of girls is common in sub-Saharan Africa and South Asia; in Niger 77% of 20 to 24-year-old women were married before 18, and in Bangladesh the figure was 65%. (*The Times* is to be congratulated on these helpful comments it inserts from time to time.)

On March 4th, 2010 the *Times* reported that a Saudi woman is to receive the lash 300 times and be imprisoned for 18 months for filing complaints against court officials and *appearing in court without a male*. The woman, Sawsan Salim, had accused local justice officials of years of harassment.

From Geneva on 5th April 2010 Navi Pillay, the UN's senior female official on Human Rights, again reported that about 5,000 women are murdered every year. Most are killed for marrying against their family's wishes or for extramarital affairs. One assumes nearly every case occurred in a country in which "honour" killings are legal. When, one wonders, are free women in other countries going to rally their forces and step forward to speak up for their sisters? One does not need much imagination to think what horrors this bald figure represents.

One has to constantly remind oneself that only the exceptional cases reach the papers in the West; there is a mass of human misery that we never hear of. One of the sensational cases was in Iran, first reported on 7th July 2010, when death by stoning was ordered for Sakineh Mohammadi Ashtiani. Mohammed Mostafaei, the lawyer who represents her, has described the sentence as absolutely illegal, based on the opinion of the judges, and that she could not follow the proceedings because she was of Azerbaijani descent and speaks Turkish and not Farsi. I believe she is still imprisoned.

Fatima, in Blackburn, England, it is reported on 10 August 2010, is now in a women's refuge. She is amazed to have met there so many other women in her situation: she refused to marry the man they had chosen. When she refused a second time, the family abused her and broke her nose. Her mother said "You are going to Pakistan to marry this guy." Fatima discovered he wanted a visa, and she was the means of getting it. She had met the man: "I didn't like him" she said, "he had a vile temper." She fled with her boy-friend in his car, and later, in the refuge, had a text message from her brother: "When I find you, I will kill you." There are an ever-growing number of women here, in Britain, like Fatima. No doubt there was money involved. Women are still struggling out of being exploited as chattels in a deal between families.

On the 11th August, 2010, a 16-year-old girl was reported as being paraded naked, sexually-abused and ritually shamed by a mob of young men in a remote part of eastern India. Sumito Murmu's ordeal would have gone unpunished if one of her attackers had not filmed it on his mobile telephone. The police saw it: "She was paraded through villages over the course of several hours" said Humayun Kabir, the District Police Superintendent, to *The Times* journalist. "In these remote villages people believe in their own law." Sumito and her father are under guard, and police are asking for donations to help her start a new life elsewhere.

In Afghanistan, under the Taleban: on August 17th 2010 two lovers were stoned to death in an execution overseen by this merciless movement; the ritual lasted more than an hour, elders and officials said. The couple were allegedly tried by Taleban militants for adultery, and were executed on Sunday, with the participation of more than 120 villagers, but the woman was eventually shot, witnesses said. A week earlier, insurgents in the west of the country whipped and shot dead a pregnant widow for adultery, and in Uruzgan province in the south a woman had her ears and nose sliced off after running away from her abusive husband.

A young mother had a hand hacked off with a machete in a fatal attack ordered by her husband after she started divorce proceedings, as reported on 19th October 2010. She was the family's breadwinner. Sher Singh, a male in the family, is alleged to have swung the machete. Both men were accused of murder.

On the 27th November 2010 there comes a report from India of a serial bigamist who is accused of marrying 60 girls and then selling them into prostitution. Their ages range from 10 to 18, and the disappearance of girls and young women had been noticed. He sold them into brothels in Poona and Mumbai, and thereby earned up to l00,000 rupees per woman. Yet only now has it been noticed, and only nine of his victims have so far been traced. And India is said to be a "democratic country"!

Another sensational case, which shocked the world, was that of a mutilated Afghan teenager whose appearance on the cover of *Time* magazine, reported in *The Times* on the 8th December 2010. A man who helped to slice off his daughter-in-law's nose more than a year ago had been arrested in southern Afghanistan, police said, in Uruzgan province. A man identified only as Suliman had confessed to holding 17-year old Bibi Aisha at gunpoint, while his son, her husband Qudratullah, wielded the knife. The girl was then dumped outside her family home and left for dead in a remote mountain village in Chora district. Then her father, Mohammadzai, carried her to a US base; she spent ten months in a secret shelter before being flown to California for reconstructive surgery. paid for by the Crossman Burn Foundation. The magazine raised concern about women's rights, and Hillary Clinton, the US Secretary of State, stated that Afghan women would not be abandoned in the peace settlement, in which the Taleban are expected to take part. Enlightenment about women's rights is not their strongest point; a Taleban commander presided over the atrocity, which took place because it was a punishment for Aisha's constant complaints about daily beatings and abuse. She was 12 years old when she was married to settle a family dispute. Women are often used in this way.

On the llth January 2011 in Iran, a woman was shot dead on a street in Tehran. This was shortly before Sakineh Mahammed Ashtiani was sentenced to death by stoning for alleged adultery, as has been mentioned. Iranian women have long led the fight for human rights for females within the Islamic Republic. One of their causes is freedom to appear without a headscarf: Nasrin Sotoudeh has been sentenced for this. "Mark my words," said Shirin Ebadi, "it will be the women who will bring democracy to Iran." We still await the day.

On January 24ᵗʰ 2011 came the news that—although India denies it—an ancient Hindu custom of sex slavery still exists in some areas in that country. It is a custom in which young girls are married to a god in childhood and then, at puberty, sold for sex—it is known as Devadasi. Yet *The Times*' reporter, Beeban Kidron, went to the "Devadasi Belt", a string of towns and villages where Karnataka meets Maharashtra in southern India. There she met Shobha; dedicated to the goddess Yellanna in childhood, she was sold in exchange for a gold necklace and 500 rupees per week to her brother-in-law. Today she runs an organisation called MASS, which campaigns against the Devadasi system, which still exists in the area. This is one of many examples of a remnant of semi-religious belief which lingers on in rural or very poor districts, because it brings in money. Women in particular suffer from these remnants.

European women are not completely free yet: on February 1ˢᵗ 2011 we read of a woman who did not get her husband's evening meal ready; she was called a whore and the housekeeping money was withdrawn. Domestic violence is an issue in Spain, high-lighted by the recent death of two women.

We return to Africa when an item on 22ⁿᵈ February 2011 announces that four officers of the Congolese Army have been found guilty of crimes against humanity and been sentenced for 20 years for leading the rape of 60 women in the town of Fizi on New Year's Day, January 1ˢᵗ 2011. The convictions, which included lesser sentences for five lower-ranked officers, marked a breakthrough in a country whose armed men have long committed extreme acts of sexual violence with impunity. Lieutenant-Colonel Kibibi Mutware, the leader of the group, was also cashiered from the army along with his co-defendants, two majors and a second lieutenant, for rape and looting which the victims described at the ten-day trial in the town of Baraka.

The mobile court of military judges and volunteer lawyers, that tried the case, was financed by the Open Society Justice Initiative of George Soros, the philanthropist billionaire, aided by the UN, international agencies and non-governmental organisations. Human rights workers welcomed the verdict, in a country that the UN has called rape capital of the world, a formal Belgian colony. Kelly D.Askin, of Open Society Justice Inititiative said: "if word about the court is spread around the country, it could have an enormous impact on deterring future crimes, now that the rule of law is finally being enforced domestically."

Lieutenant-Colonel Kibibi and his soldiers were more than a little stunned to find themselves on trial before this ground-breaking domestic mobile court. One woman witness said: "If we go to the river for water, we get raped; if we go to the fields for food, we get raped; if we go to the market to sell our goods, we get raped. There is no peace." Thanks to George Soros and the others, perhaps they will at least be spared that in future.

Laura Logan, an American woman journalist, was attacked by men in Egypt, on 30th April 2011. "My clothes were ripped to pieces, and for an extended period they raped me with their hands", but she was eventually rescued by Egyptian soldiers when alerted by a group of women. She was flown out of Egypt by chartered plane, but decided to talk about her ordeal to alert women journalists to the dangers of such behaviour in countries where women were treated badly.

In Senegal on the 29th of November, seven hundred villages have declared an end to female circumcision and forced marriages, after 3,000 people called for an end to the practice. About 4,500 communities have been working with communities to end this barbaric custom, but in the village of Kolda 98% of girls are still subjected to it and forced into early marriages.

In Afghanistan on the 8th December a journalist, Jerome Starkey, sends in a report from Kabul to the effect that a man has been arrested for helping to slice off his daughter-in-law's nose and ears more than a year before. This is the case of Bibi Aisha's mutilated face, already mentioned, which caused outrage when it was published on the covers of *Time* magazine. Police in Uruzgan province said that a man, identified only as Suliman, had confessed to holding the teenage girl at gunpoint, while his son wielded the knife, in 2009. It transpired that a Taleban commander presided over the brutal punishment, which was masterminded by Suliman and orchestrated by the husband, Qudratullah, who used the knife, as a punishment for Aisha's constant complaints of daily beating and abuse. She was 12 at the time of the marriage and 17 when, after the atrocity, she was dumped outside her family home and left for dead. Her father, Mohammadzai, carried her to a USA base, and later she was flown to California for reconstructive surgery. There is talk of a "blood-feud" as a result between the two families, but it would be preferable in this day and age, as the magazine *Time* suggested, that women's rights were built into Nato's withdrawal.

This is but a fraction of the total of the atrocities carried out upon women in the East, and yet their sisters in the West remain concentrated on their own nations. I have complained about this to the Fawcett Society, of which I was a member, but they plead shortage of funds as an excuse for their inaction. Yet an e-mail, in these days of the world-wide web, is free.

I do not understand how it is possible for a world like the present, yet we can apparently coolly read about atrocities like the above in the daily press without a tremor. This is unacceptable, and I think it only comes about because the world is run by men. If women throughout the planet were of equal status these matters would be acted upon at once. I do not think that it would be any better if it were run by women alone; I am not that sort of feminist: it would be equally bad, if different. All I ask is equality.

Channel 4 on TV announced on 22nd August 2011 that women in Rumania were so troubled by poverty that they were compelled to seek prostitution.

Even as I write this, the mail drops through the door: it includes an appeal from UNICEF, the United Nations Department that seeks aid for children—they are sleeping in the streets. This would not happen in a world in which women were equal with men in all countries.

Some time ago the President of Afghanistan, Hamid Karzai, signed a law which, in effect, allowed men to rape their wives and/or starve them.

And, finally, *The Times* on 1st March 2011, in a tiny item, announced that the former Argentine dictators, Jorge Videla and Reynaldo Bignone, appeared in court on charges linked to the kidnapping of 500 babies taken from their mothers from 1976 to 1983. What does this mean? We shall, I hope, find out.

I believe that at one time, many thousands of years ago, women were equal with men everywhere. In hunter-gatherer societies, they gathered and the men hunted. When decisions were made by the tribe, the Elders met and made the policies, men and women alike, I assume. If there were any fools of either sex, nobody paid any attention to them: they were not consulted. Everyone knew who the most wise were: by common agreement they were invited to make the decisions. They decided at what time the tribe would move up to the hills where the pastures were just approaching the right moment; they decided when they would return, and so on. When there was disagreement: say over a hunting-ground, they met and came to a conclusion about it which both found convenient. How all this came to an end we do not know; perhaps, one day, we shall.

I have great faith in the women of Egypt. We have already seen that they had a culture once in which they were very strong. I do not know when the Fall came, but I am sure that they fought it; in every culture in the world the property went through the female line at some time in their history: it was universal originally, for obvious reasons, and still is in some countries.

Years ago I read a book by a woman American anthropologist who had discovered how certain tribes dealt with discord within the tribe: the tribe slept on it, and the women dreamed a dream. The next day the tribe gathered, the women sang their dream, and the men danced to it. Meanwhile the wise men passed among the people, talking to the main protagonists.

Eventually, the women ceased to sing, and the men protested, but the women had finished. It was all over. Everybody went to bed, and that was the end of it. Now I know that it is important, I have been trying to find since who was the author, but without success. I think that was a true picture of how a tribe lived in the time of the hunter-gatherers.

Why is this cruelty to women permitted in society today in South America, all of Africa, and most of Asia? I think it is because the men fear them. If only men would learn to accept them as equals, their reward would be great in the long term. The whole of society would be healed. Half the wisdom of the world would be released, and the world would be the better. These unfortunate women are the equal of their kind in the rest of the world; they are missing out on the ability to show their intellect, and who knows what poets, scientists, and geniuses the world has lost. There is resistance to the little that is being done. *The Times* reports that warlords who hate women are behind Afghan government plans to take control of women's shelters. Human Rights Watch said the move was a sop to Islamic hardliners, such as the Taleban. The shelters, which protect hundreds of women from forced marriage and ritual violence, are mostly run by charities and the United Nations, while the Taleban is a movement that is particularly against freedom for women; any agreement which is made with them must be scrutinised with care.

On the 14th March 2012 *The Times* reported that Samira Ibrahim had announced that she was going to take her case to an international court after an Egyptian court acquitted the defendant of all charges. She was one of seven women allegedly forced to undergo "virginity test" while in detention. "I have done with Egyptian Courts," she said, "I will go to the international courts for justice". I salute this brave woman, and wish that women everywhere would join in with her.

Chapter 3

There is a new World Coming Into Being

One of the oldest sayings is in the Old Testament of the Bible: "there is nothing new under the sun". No longer. Now, when something happens on the Antipodes, it is known on the opposite side of the planet within five minutes. We have the Internet, and now Facebook, Twitter, etc. No longer can any dictator take over and decide to have his way by mere show of force, he (it is always a man who behaves thus) will be questioned by the new medium. For the first time in the history of the world, we are conscious of the rest of the planet.

Everything has changed—except the human mind. We have become so used to the world as it was that we cannot comprehend what has occurred; we still carry on behaving as if it were unchanged. Already there are ripples on the surface, but we look at them and talk of an Arab Spring. It is much more than that: it heralds a change to the life on the whole planet, plus a lot of cultures; we must also change, and fast.

It is necessary to re-emphasize what I have already stated: this is so important that it must be laid before you simply, in words which every person will understand, man and woman, not in academic language. We are human beings, of different cultures, which must be critically surveyed, now living on one Earth, and if we do not rapidly comprehend this fact the layer of life which has developed upon the whole unique planet, and the creatures living within it, including ourselves, will suffer and die. What a unforgivable waste!

Accept that life has changed, we are now living on one planet, not only in one country, we have entered a new chapter, leaving behind us the old world, where we embraced our individual cultures, in their different forms. All of them are, to a greater or lesser degree, man-made and imaginary.

An example of this is the belief in this life being the only one we know. Once it was taken for granted that we lived many lives; this was assumed in every culture, but it was changed at a key time, I discovered, by members of the Christian faith, who were at the time competing with Buddhists for members. As reincarnation plays a large part in the Buddhist scheme of things, the Christians decided to abandon it: they declared that there is only the one life on earth, during which we commit many sins, but if we believe Christianity we shall be forgiven all of them, through the sacrifice of Jesus, who died for us on the cross.

Unfortunately I have no reference for this Christian decision, because the work of reference, which held it, was borrowed from me and has disappeared. Nevertheless I have discovered no other reason for the difference. It is a great loss, because before it disappeared, this concept was a salutary thought for the rich elite: that they might be born a beggar in the next life.

I also watched on television recently an item which gave me deep satisfaction: it confirmed my suspicion that some animals had developed their brains a lot more than had been assumed by us. It was presented by Dr Liz Bonnin, who approaches the subject scientifically. She proved that this is indeed the case, and that it applies to what might be called the higher mammals: elephants, whales, dolphins and the like. She shared with us her obvious joy at the offspring of a small species of whale, when the mother, again and again, encouraged it to return to be stroked by her human hands.

She also demonstrated that they communicate with each other. We saw horses, elephants, coyotes, and monkeys do this by movements of the body. These animals also can understand our motivation in simple matters such as hiding food, in which they have an interest. This particularly applies when they are group animals, and bond within the group. The bond is illustrated by a finger to the eye—understood, too, by humans!

We watched elephants playing football with human beings. Also, a difficult problem was set for the same animals to solve; between them, after one or two tries, they used communication between them to find the solution. Such information will not be a surprise to dog-lovers, who often are impressed by their pet's sagacity. Dogs have worked with human beings longer than any other creatures, and have learnt to read our minds. Some species of birds, notably corvids, are very clever at discovering human habits with regard to food.

This research opens up a new world of communication to us, and new friends. Dr Bonnin and her team also devised clever, rather sinister, masks and then carried out certain irritating actions, and saw whether the birds remembered. They certainly did, and tended to mob the wearers when they returned. These memories they remembered for long periods; we humans have for many generations remarked that "elephants never forget", and it is true.

Reincarnation was a world-wide belief long before the present. In the Bible, it was taken for granted: Jesus asked the disciples a question: "Whom do men say that I am?" They answer with the names of various people, and he presses them: "But whom say ye that I am?" And Simon Peter answered: "Thou art the Christ. the Son of the Living God", which was the answer Christ wished. (Matthew chap. 16 v.13).

*In *The Tale of Prince Genji*, said to be the first novel ever written, in Japan, and a thousand years old, one of the characters remarks, early on: "It must have been something relating to my previous life".

And in Hugh Brody's record of his time with the Inuits of Canada, *The Other Side of Eden*, H*unter-Gatherers, Farmers and the Shaping of the*

---

* ISBN 0-571-20596-8

World,[*] he writes: "parents and grandparents look at every detail of a newborn to see who has returned: there will be a birthmark, to tell them which loved one they are welcoming back."

All these things: the internet, the passing of many religions and the realisation of the rightness of the most ancient of all—have perhaps affected me so powerfully because of my generation. Born in the last years of the First World War, and slowly realising that humanity had just passed through something so terrible that it could hardly be expressed, though books were to be attempted, as time passed: *All Quiet on the Western Front*, and many memoirs, such as that of Robert Graves, gradually mounted up, and those of us who read them realised what horrors we had survived, and then had to come to terms ourselves with a world that was changing very rapidly, and continued to change. I speak as one who has lived longer than most, and find it difficult to keep up. Nevertheless, I must speak, for it appears that nobody else is going to do so.

We can have a world where experts will talk to animals, as witness Dr Donnin's work, where everybody (and in the poor countries, everybody who is rich—because, curiously enough, there are people who are well-off there) has mobile phones, and the planet will be joined for the first time by everybody else who dwells on it. The only barriers will be language. These will be overcome rapidly, especially with the coming of education for all. Where there will be enough room for all, especially in Britain and the Maldives, who are

---

[*]   Publisher Faber and Faber, 2000

suffering—already in Britain because she is already crowded, in the Maldives because the people have been pushed out, by the sea.

We can, we must, start at once on the things we can do. Reduce the human population: for this we shall need the women. First those who are sufficiently free from their own culture can ask themselves, (and their partners if they are married): do I want to become a mother (or a parent) or to be free for ever to make my career my purpose in life. If they have a husband he will have to decide, too.

Chapter 4

The History of Humanity

A recent British Education Minister stated during his time in office that the children of today were starting too early on history in our schools: they should start at the time of the Renaissance. I gazed at this statement with amazement. If one is going to understand anything, one *must* learn how it came about. To do this, one must begin in geological time: archaeological science tells us that humanity began in the time of the hunter-gatherers, some 195,000 years ago. The men hunted and the women, young people and the old and frail, gathered; these tasks required a considerable depth of knowledge. I would like Michael Gove, the Education Minister in question, to sit down with a sizeable tome, which is among my books, and read it. It is called *Eden In the East*, by Stephen

Oppenheimer[*], and is a meticulously detailed account of three great floods, which have occurred in South-East Asia in prehistoric times, and the events which were thus precipitated, along with the effects among the disturbed people. (If only South America had been similarly observed!) Dr Oppenheimer is basically a medical man, and his curiosity was aroused through his observations, in the course of studying *thalassaemia*, of the drifting of folklore around the area, sometimes over thousands of miles, such as the stories of Cain and Abel and Noah's Ark. These stories get garbled and the names differ, but they are recognisable.

He writes of the development of farming among the hunter-gatherers, and I have always thought that it must have been the women who first recognised the effect of animal manure upon plants, and then experimented, and this in turn led to the development of farming, which in turn resulted in further specialisation, so that gradually the concept of different sorts of communities began: villages with a blacksmith, shoeing horses; a woodworker making furniture; a wise-woman collecting herbs which healed certain maladies, and so on. But all this occurred over millions of years, and moreover on another plane something else developed: religion. And both developed in one part of the world at first: the continent of Africa.

In the present state of the science of archaeology we have not discovered much of this stage yet: the curiosity about the prehistory of the different stages,

---

[*]   Publisher The Orion Publishing Group Limited.
ISBN 0-7538-0679-7

as human beings developed, seems to have been quite racist; such people have not apparently been regarded as worthy of interest. Only in Europe has there been any consistent research archaeologically, and war has often eliminated certain valuable buildings elsewhere before they could be examined—the so-called civilisations of today have departed from the peaceful habits of the early religion described in the *Introduction*; nowadays we spend large parts of our wealth on weapons of war.

The development of *Homo* began with the Bushmen of South Africa: small people with peaceful habits until larger races developed and began to persecute them:[*] The brown Bushmen are described by Sandy Gall: "They came early in the morning, four of them, stepping lightly through the bush, as insubstantial as shadows against the sun. One carried a bow and a quiverful of arrows, each of the others had a spear . . . . They were dressed only in skins, a loincloth made of antelope hide, probably springbok or gemsbok, and their torsos, legs, and even their feet were bare. Each man carried a little pouch . . . . Bushmen have, and need, the minimum. Small and lightly muscled, they move with an oiled activity altogether delightful to watch."

After the Bushmen, black races developed, full sized and—unlike the Bushmen—spread all over the world. When they stayed in temperate regions, they lost their blackness and in the North became blond.

---

[*] From 'The Bushmen of South Africa', by Sandy Gall. ISBN 0-7126-8436-X

One more occurrence in geological time was the development of religion in humanity. Forever curious, humans asked themselves how to explain life, and, they assumed, gods did all the inexplicable things. The importance of the crops took priority above all others, because if they failed it was a calamity, and they thought the gods were angry. They appointed a female goddess—the Greeks called her Demeter—to look after fertility, and when humans realised the connection between the act of coition and birth, the custom arose of choosing a human male to be the representative of the growth of the plants they sowed. Nothing was too good for the Chosen One during the time of his reign, he could choose any maiden he wanted, every wish would be granted, and he had to be physically perfect.

We know something of these young men, because their reign ended at the harvest, when they drank a draught of stupefying herbs and were sacrificed and the blood was sprinkled over the crops, and then were buried ceremoniously often in peat which kept them in the same perfect physical condition, so that our archaeologists, millions of years later, could examine them. We have some 85 of them, Bog People they are called: their nails are perfectly manicured and everything about them is preserved in the condition in which they died.

There is also a connection between this Chosen One and Christianity: it is not difficult for the concept of the one who gives his life for the good of the community, and dies for it, to be transferred to a new religion. It is possible that this occurred; only the science of archaeology will tell us.

This is the story for 40 thousand years, then we change to human time: hours, minutes and years. The Earth is populated, but thinly; men and women have not yet become aware of the world as an entity. But the human being is endlessly curious, and the Vikings seem to have penetrated even into America and left traces of their stay.

We can only wait until archaeology tells us the details of the terrible transformation from peaceful development into strife and the degradation of women, which came later. We only know that the fabled Golden Age ended, the concept of a male god instead of the Great Goddess took over, together with rape and brute force—perhaps that was when sex became shameful.

Humanity is still waiting for archaeology to tell us how this occurred in detail, but we know something of how it happened: the season of spring—as indeed, everything which we did not understand—was the work of a certain divinity: the word Easter comes from Oestre, the Greek goddess of that season, when in the temperate climates the weather undergoes the familiar annual change, the birds begin nesting and buds burst open on the trees. (She is only remembered these days in medicine: some imaginative and learned chemist named the drug *oestrogen* after her.) We also still await the details of how the peaceful calm of the farming period was broken by the irruption of the Northern gods, mentioned by Bradford, with their thunder and lightning, and how they took over, and in which regions. Some archaeology has discovered a certain amount of the answer even as I write this: the historian Bettany Hughes, on the BBC, has described digs which have recently started in various places on work which reveals a story of a Magna Mater (Great Goddess), under various names.

She told us that in the ancient catacombs of Rome there are portraits of women taking part in unknown Christian rituals, and how in other parts of the world priestesses have continued to play an important role. Then something happened—perhaps archaeology will tell us the details in time: Ernle Bradford says, as we have seen earlier, that male Northern gods appeared, wielding thunderbolts, and the dominating culture changed, heralding a sad period in the history of humanity: a period in which cleverness and love of profit took over from peace and wisdom—resulting in the Warrior Age, and patriarchy.

However, everything and every creature on Earth depends still upon the ecological web, whether human or not. Since those days we humans have done tremendous damage to the web: from the great top predators to the little creatures of which we do not know the names, they are so obscure, but so vital. For example I noticed the other day that a fox had shat on the stones outside my door to the garden. Next day this had partly been consumed by some tiny creature, and the following day it had disappeared altogether. What insect was this, which usefully consumes every bit of excrement which is lying about? I looked at my reference books and discovered to my surprise that there are a number which do this useful service. But all are vital and all are threatened.

The oceans are becoming acid, its shell-forming creatures soon will be unable to build their homes, and human activity is solely responsible. The precious oceans in which life began on Earth, because it possessed water, have been polluted, not by some wicked divinity—we have stopped blaming the gods—but by *Homo sapiens* himself, not so sapient.

At the present stage of history men are still running things, but we might have to change our system: capitalism is perhaps acceptable so long as the values of ecology dominate, but at present they don't. Wise economists, of both sexes, can decide in the future whether we allow ourselves to continue to choose capitalism as our system, with its "profit at all costs" motivation. Meanwhile, we have nearly passed the point of no return—some say it is already too late—the ecological web is almost irreparably damaged, and the gene pools of the great predators are almost too small for them to continue.

Yet nothing at all is being done about it!

This is astounding: surely some of our wise elders have observed this calamity? Yes, of course they have; but immediately someone starts to do anything about it, it fades out. If we look into the reason for this, it is because it is unprofitable: not in the interests of some huge corporation. For example, why are manufacturers still allowed to make goods, which cannot possibly be recycled? Instead, we now send the waste to a poorer country to be disposed of. There is something wrong with a community that acts like this. The only excuse is that we have not woken up, really, to the new situation. We have got into the habit of only thinking—if indeed we think about such things at all—about our own country. The time has passed when that sufficed. Now we have to think of all humanity on the planet, and what we are going to do about the ecological web on which we are, in the long run, totally dependent.

First of all, I think we must go more deeply into this subject of why it is fading out: because it is not profitable: profit now rules the world. For example: let us look at the latest idea: The Stratospheric Particle Injection for Climatic Engineering—SPICE for short. It is an idea, backed by a £1.6 million British Government grant, of a helium balloon, with a diameter of around 150 metres, which would allow scientists to pump sulphate or other chemicals safely into the stratosphere and reflect back into space a fraction of the incoming sunlight. This invention is inspired by the cooling of global climate observed after the eruption of Mount Pinatubo in the Philippines in June 1991, which reduced global temperature by about 0.5C for two years. Such a balloon would, of course, be highly profitable for some great corporation to make.

Perhaps this is the moment to interpolate something about the institution of the Monarchy in Britain, but only after beheading one of its representatives and later depriving the others of the privilege of the vote, thus rendering it tame, and impotent politically. The Queen's Speech, at the beginning of a new Parliament, is written solely by the Prime Minister of an elected Government. Other dynasties, less dramatically, among the monarchs of Europe, have eluded their end, and indeed are regarded with affection. This is because they have all renounced dictatorship and taken on the traditional role, which is of immense value: it is only revealed in the tremendous curiosity and lasting interest taken in these few remaining Royal Houses by nations which have lost theirs: an example of ancient wisdom.

I would just like to add something of my own: I have a feeling that we are still of the Tribe of the Horse (from our Celtic tradition). Hengist and Horsa mean, I have learnt, mare and horse, and I have had one volume among my own books, which traces back the Royal House to the Priest of Woden. The affection of Her Majesty for horses is well known; I rest my case.

We Realists want to spend time on reducing the polluting of the Web by the human race, *immediately*, not tickling the subject with a balloon which might reduce a part of the global warming for 0.5C for two years. Tackling the primary task, over-population, will take years, it will bring in no profit, but it is the only sane thing to do. Every environmentalist is hopping mad with this insane balloon suggestion, but our politicians, motivated by the capitalist values, obviously think differently. Short-term, of course, and the masculine love of scientific toys, both overcome the need to cure global warming, despite the urgency of the task; no balance by female values, *long-term* and not *combative* nor fascinated by toys.

Taking one subject as an example—water consumption in Europe. We use so much in our ordinary domestic life now, that we are affecting our rivers; once beautiful and filled with wildlife of all sorts, especially in chalk areas. Soon after leaving the spring they become affected by the amount of water drawn off by our consumption, and then what life is present is starved and poor. A quarter of our rivers in Britain are starved in this way. A part of one town, South Swindon, according to the BBC doing a survey on the 19th September 2011, takes from the River Kennet 150 litres every day. Yet again, there is a short-term attitude to the problem: instead of doing something about the root cause and encouraging people to save water until the population is lowered (and the BBC did the public a service by drawing attention to the matter), the powerful corporations in charge of water "pour concrete" and make money out of the work: they want to build reservoirs. By doing so they would take even more land out of the little we have. At present the BBC reckoned in this programme that it would take £17 million at present to bring back the River Kennet to pristine condition. Apart from domestic use, water is drawn off by farmers irrigating their crops, and industries take a heavy toll. And that's just one example among thousands.

In short, the human race is the most polluting species which has ever developed on Planet Earth. We have the cunning to overcome all of our major predators, so there is nothing now to hold our population back at a reasonable number. We pollute everything that comes into our spheres of action, and they are many. From the Earth itself, which we fill with our waste-pits, to space, which we pollute with bags of urine on the chaste Moon, to our rubbish which circulates round the Sun in ever-increasing amounts. This has to be done at the same time as we deal with the surplus population and the end of the degradation of women, but it *must* be given equal attention; we have much to do.

A speck of matter, going at high speed, is already an object of fear to anyone entering space. One item in our future agenda might be the creation of a new industry: beginning the mining of old landfills for recyclable items, and burning the rest for heat.

Why are we not pressing on with exploiting energy from the Sahara, and elsewhere in tropical countries, by exploiting whole tracts of desert by using masses of solar panels collecting it? We can pay them for it, thus enhancing their income. Why are we not improving sailing ships by modern design, which is more efficient, incorporating solar panels, instead of fossil fuels, when there is no wind? Why are we still spending money on making armaments instead of banning nuclear weapons and war as a means of solving disagreement among nations?

I dream of a clean Earth with a reduced human population, then Nature can breathe a sigh of relief. We can begin creating a civilised life right now upon our exceptional planet. Everybody can be fed well, will have enough water, sanitation, and shelter and education: we shall be able, at last, to create the good life for all. We can learn how to communicate with the higher animals. It's up to us. Our values have to change to long-term, feminine ones: we *must* not rest until no one on this planet is starving. And we can begin at once upon the control of human birth.

I shall never see the day when it comes, but I imagine the joy of a planet when it can look around and see space enough to watch the wildlife, and wildlife enough, with gene pools big enough, not to make us worry.

Simultaneously, we shall work to free our fellow-women, until they can all enjoy the freedom outlined in the marriage agreement quoted above. This will take time, but it will be worth it for the happiness it will bring to both men and women. Once men have taken aboard the notion of women as comrades at their side, and not the shrews and bullies they fear, they will enjoy the freedom of the creativity they can use—completed creativity; we have seen it once or twice in the world, where a man and a woman have worked together to build some enterprise. Perhaps it will come first in the struggle for peace, where teams of men and women come together to staunch the flow of blood which results from the factories of weapons, and save the expenditure to put it to good work helping improve the healing in free medicine in the new Health Services of the world.

Smaller matters, too, we can deal with at last. We shall make our prisons places of healing, where people can get over bad parenting or bereavement in childhood. They will be in rural surroundings, with gardens and animals and farm work, even if this is inconvenient to visitors: they will have free passes. New sorts of books will appear: *Skimming Stones* is the first, by Rob Cowen and Leo Critchley,* which approaches Nature in the new way. *The Good Earth*, by Pearl Buck, which describes a different culture from the eyes of an American, but with a depth of understanding very rare** in the present world.

And I have to admit that ever since George Bernard Shaw was failed by the British Museum in carrying out the Shaw legacy to modernise English spelling, I have wished to try again.

We have become so used to being crowded in Britain that it is difficult to envisage the world as it

---

*   ISBN 978-1-444-73598-7

**   ISBN 978-1-4165-1135-9

will be. I wonder if we shall decide to do what has been suggested: build an underground world, save for the most beautiful cities, which will be preserved and cared for, and some little old buildings, just to remind us of what they were like. Then we shall live snug underground, free forever from the vicissitudes of weather, but people will go out and enjoy exercise, running, sports, ball games, hunting, in the old ways—almost forgotten, as described in *Skimming Stones*, from time to time, enjoying the sunshine when we choose to, and let the wild things take over once again. We shall learn how to speak with them and enjoy their company. The country will once more become a place where we shall have to proceed with care; we shall know the old, wild past of humanity. With our numbers down to normal, we shall be in balance with Nature.

We should spend more time dreaming of life as it might be, and forget life as we see it changing before our eyes into something monstrous. Grasp the future: a new freedom can truly be yours, if you will just make the effort.

I am, at the last moment, given encouragement by hearing a programme on the BBC from a historian, Bettany Hughes, about the archaeological discovery that in fact in prehistoric times, 40,000 years ago, humanity worshipped a Great Goddess, a Magna Mater. Even more interesting was the final piece of my jigsaw, which arrived with the monthly scientific supplement of *The Times* on the 3rd May, 2012: *Eureka*, completing the picture; it contained an article by Michael Hanlon on the completion of the Shard building, and was filled with information. Like most people, I was unaware of the Green effect of such a building: once a residential building exceeds about 40 storeys its apartments begins to exceed in intrinsic value the cost of construction, because the "footprint" is shared between many living units. Continue to build, and the land costs become almost irrelevant. (Edward Glaeser, the US economist, writes in his recent book *The Triumph of the City:* "Canyons of glass and steel and concrete, such as those along New York's Fifth Avenue, aren't an urban problem: they are a perfectly reasonable way to fit a large amount of people and commerce on a small amount of land." Moreover, building up keeps homes affordable. The most vehement opposition to vertical expansion usually comes from wealthy householders in up-market low-rise areas, such as Greenwich Village in New York or Chelsea in London. Part of their objection is aesthetic, but it also comes from the knowledge that by driving down property costs, the exclusivity of their neighbourhood will be diluted. Tall buildings are Green. New York or Los Angeles, and other mostly low-rise cities, such as Houston or London, are ecological nightmares: by building low they lose all the economies of scale that should be inherent in a city. Low buildings, per unit volume and inhabitant, consume more energy to heat and keep cool than tall ones. The lower the city the more it costs to move water, waste, power and sewage to where it needs to be. Recycling glass, paper and plastic makes sense in New York but less so in Los Angeles, where the energy cost of collecting the stuff can exceed the energy cost of recycling.

High-rise environments are much easier to get around. Because land use is so efficient, commutes are typically short and public transport networks can be integrated with centres of business and housing, burn more calories and bicycle more. In the US, obesity levels correlate negatively not only with the degree of urbanisation in a given area, but also with housing density and the height of buildings. The thinnest urban Americans are New Yorkers, the fattest are the denizens of the sunbelt sprawls.

As I have stated above, the human population on this planet has reached a size when it has become a burden to the system, passing now a figure of seven billion, and in 2010 we became an urban species, with more than half of us living in towns and cities, in which building techniques too are changing: traditional skyscrapers in the 20th century were essentially steel-framed concrete boxes, all looking the same, and stresses from wind exceeded those from weight. At this stage I must disagree with the writer of the article, who goes on to state "the city is arguably humankind's greatest achievement". It is not. Our pullulation, like that of rabbits, and our intelligence, have combined to enable us to overcome our predators, apart from certain bacteria, and to have a negative effect on the ecology of the planet.

The building of The Shard has introduced new techniques: framed-tube construction, with the tube on the outside, acting like a cantilevered bridge; the stresses from wind and weight born by a series of columns and beams. Tubes can be bundled together to provide mutual support, allowing one central (or off-centre) tube to reach a great height. The needle of The Shard, built by Mace construction for the Sellar property group, relies on a single simple concrete core to provide local load-bearing strength. I watched on the BBC the speedy and proficient erection of the building with admiration.

But these advantages of high-rise buildings are only applicable if we put the other matters right: reduce the population, raise women to equality throughout the world, reduce poverty and raise the standard of living, and stop using war as the solution of problems, which it never is, and spend the huge amount of money it costs on the reduction of poverty. Above all: *Imagine*, as the Beatles once wisely suggested.

This may well seem an impossible dream. Men have often said that of a female suggestion, but in fact it is pure common sense. It may seem to be impracticable, but it is the *only* practical, pragmatic way ahead. And in the long run it will prove hugely profitable in the female way—to human beings in the things which matter: it will bring true contentment and peace.

www.ingramcontent.com/pod-product-compliance
Lightning Source LLC
Chambersburg PA
CBHW021244280526
45784CB00005B/2235